Deep Space Cubesats

Patrick H. Stakem

(c) 2020, 2022

Number 8 in the Cubesat Series

Table of Contents

Introduction

This book covers the topic of Deep Space Mission, using Cubesats. By Deep Space, we mean Jupiter and beyond, but we will present a case for the Asteroid Belt as well. The core concept here is, there is strength in numbers. Rather than sending one or two explorers, we will suggest sending 1,000.

NASA, particularly Jet Propulsion Lab, is defining new approaches to exploration away from Earth's neighborhood, utilizing Cubesat technology. Many missions using solar sails are being considered, and the technology has been demonstrated. Up to this point, the application of Cubesats for interplanetary exploration has been approached by building bigger, more robust Cubesats. Later, we will discuss another approach involving standard launch vehicles, carrying large numbers of Cubesats.

Author

The author The author has a BSEE in Electrical Engineering from Carnegie-Mellon University, and Masters Degrees in Applied Physics and Computer Science from the Johns Hopkins University. During a career as a NASA support contractor from 1971 to 2013, he worked at all of the NASA Centers. He served as a mentor for the NASA/GSFC Summer Robotics Engineering Boot Camp at GSFC for 2 years. He taught Embedded Systems for the Johns Hopkins University, Engineering for Professionals Program, and has done several summer Cubesat Programs at the undergraduate and graduate level.

Mr. Stakem has been affiliated with the Whiting School of Engineering of the Johns Hopkins University since 2007. Mr. Stakem supported the Summer Engineering Bootcamp Projects at Goddard Space Flight Center for 2 years.

Mr. Stakem can be found on Facebook and LinkedIn. Comments, corrections, suggestions are appreciated.

Cubesats

Mars Cube One (MarCO) is the first interplanetary cubesat mission, headed by JPL. It involves sending two 6U Cubesats to Mars, along with the Insight Rover. The 6U cubesats separated at Earth orbit and proceed on their own. The mission launched in May,of 2018.

The Cubesats will serve as a real-time communications relay with Earth during the critical descent and landing phase of the rover. The lander talks to the Cubesat relays over an 8kbps UHF link, and the Cubesats relay this to Earth over an 8kbps X-band link to the DSN.

The Cubesats are stabilized with reaction wheels, and have propulsion systems to unload the wheels, and adjust their orbital position.

Where are we going?

For purposes of this book, I am discussing destinations beyond Neptune. Jupiter and Saturn get all of the attention, but there are even more interesting things further out.

And, actually, more interesting things at the same distance. Trojan asteroids are in the same orbit as their primary planet, but spaced out around the orbit. Some of these asteroids are the size of moons. The Japanese mission OKEANOS is being developed to study some of Jupiter's Trojans. It uses a hybrid solar sail with solar panels to power an ion engine. It is a conceptional spacecraft, not a cubesat, bur the concept is applicable. One of its concepts is to send a lander to the surface, to collect a sample to return to Earth. The spacecraft will have an onboard mass spectrometer. They are including payload processing onboard. They expect their approach will reduce the cost of deep space exploration by an order of magnitude.

Pluto

Pluto was downgraded from a planet to a Kuiper Belt object. The *New Horizons* mission to Pluto and the Kuiper Belt began in January of 2006, and reached the vicinity of Pluto in July 2015. It conducted a 6-month survey of Pluto, and went out farther into the Kuiper belt, on an 3 year extended mission, which is ongoing at this writing. The spacecraft was developed for NASA by the Johns Hopkins University Applied Physics Lab in Laurel, MD.

In March of 2007, the mission computer experienced an un-correctable memory error and rebooted itself, causing the spacecraft to go into safe mode. The craft fully recovered within two days, with some data loss on Jupiter's magneto-tail. The one-way light time back to Earth was 4.6 hours.

In 2015, the Pluto flyby occurred, and data began to flow back to Earth. It took more than a year for all the imaging data to be transmitted, due to distances and transmit power involved.

Pluto had one known moon, Charon, before the New Horizons Team members, using Hubble Space Telescope data, discovered four more, Nix, Hydra, Styx, and Kerebos.

Comets

A comet is an icy solar system body usually in a large, elliptical orbit. There are 5,253 known. As a comet approaches the Sun, it heats up and boils off material that forms a long tail called a coma, it's most spectacular aspect.

Not everything in our solar system is from around here. Generally, our solar system bodies orbit the Sun in a disk called the ecliptic plane. Comets that are not necessarily in orbit around our Sun can take a path that are highly inclined to that plane. As this book was being written, the first asteroid at a very high angle with respect to the ecliptic was observed. That means it did not originate in our solar system, but came from some where else in our galaxy, or beyond. There is no particular name for this class of objects, but the title "Exeroid" has been suggested. This will have to be cleared with the International Astronomical Union. The object was observed by a telescope on the Hawaiian mountain of Hakeakala. It is the first time an interstellar object has been observed. It was named Oumuamua, in Hawaiian, "messenger from a far, arriving first." It came from the direction of the constellation Lyra. It was first spotted on October 19, 2017.

The Diminutive Asteroid Visitor using Ion Drive (DAVID) is a 6U Cubesat mission to visit an Earth orbit crossing asteroid. This is a project of NASA-Glenn.

Kuiper Belt Objects

The Kuiper Belt extends from the orbit of Neptune out approximately 50 AU. There are three known dwarf planets, the formal Planet Pluto and two others. Over 100,000 units are speculated to exist. Neptune has a major gravitational influence over the Kuiper belt objects. Not much is known about the belt and its objects, since astronomers have had to rely on ground based observation. The New Horizons mission is proceeding out through the Kuiper belt, and will tell us what it sees.

Beyond the Kuiper belt is the Scattered disc, extending beyond 100 AU. This is a sparsely populated region of the solar system. This is no knowledge of how many scattered disk objects (SDO's) exist. The closest is at around 30 AU. The belt extends above and below the ecliptic plane, in a torus configuration.

Beyond this is the Oort Cloud, extending out 5,000 – 100,000 AU. There is a disk-shaped inner cloud, and a spherical outer cloud. And you thought space was empty.

I didn't total all this up, but it doesn't seem like this is a small task. We need to get a lot of spacecraft working on it. U.S. Spacecraft have visited all of the planets in the Solar System (including the demoted Pluto), and many of the other objects as well

Interstellar Space

You want to talk about interstellar Cubesats now? Actually, Cubesats can play a role in Astrophysics and space astronomy by observations of other star systems. In space, particularly behind the Moon, when the Sun's light is blocked, is good good observation point. The L2 point is the destination for the James Webb Space Telescope. An approach that Cubesats can provide is distributed observation, using a station-keeping swarm to implement a large synthetic aperture telescope.

A proposed project to understand the energy transport from black holes would observe in the 5 MHz band. This has to be done away from the Earth's "radio pollution." They postulate a constellation of 100 Cubesats in a 1 kilometer diameter sphere, with 10 cm station-keeping. There is a mothership that serves as the constellation's communication relay, and pre-processes the data to control downlink bandwidth.

Other Stars

We would like to visit other star systems to see if Earth is unique in hosting life. The closest star to our Sun is Proxima Centauri in the southern constellation if Centaurus. It takes more than 4 years, traveling at the speed of light, to get there. Of course, we have no clue on how to accelerate to the speed of light. It may not even be possible. For a Cubesat mission, we might take 25-50 years to get there, with the small constant push on the solar sails, supplemented by the ion engine. Actually, we have to turn around and have Proxima Centaurus slow us down, otherwise we may just zip past it at high speed. We'll probably run out of battery power before then. We have to worry about the electronics lasting. One concept that has been discussed it to build a large, solar-powered laser in the outer reaches of our solar system, and use that, targeted directly on the explorer's solar sail. This might speed up the mission a bit, but we still have to slow down at the destination.

With multiple explorers, we could dedicate one to the star, and others to any exoplanets we might encounter.

The other issue is communications, We have to transmit the data with enough power to be received in a usable fashion at Earth. This would probably involve a laser on the explorer, and a dedicated telescope, in orbit or on the moon as a receiver.

There are significant challenges to visiting our nearest neighbor, and we really don't want to wait that long to get some answers. As technology improves we might, for example, launch a faster vehicle that will overtake our first probe. We better get working on those interstellar Cubesats.

How are we getting there?

In 2015, the Planetary Society's LightSail-1 successfully deployed its solar sail. This was done in Earth orbit. Planned follow-on projects LightSail-2, 3, and 4 will follow. Lightsail-2 will be launched shortly after this book goes to press, with 32 square meters of sail, and advanced guidance electronics. It is a 3U Cubesat. It will deploy its sail at 800km. LightSat's 3 and 4 will be more than technology demonstration, with the 4th unit heading to the L1 Lagrange point, to provide earlier warning of Solar geomagnetic storms. Other systems have been proposed with continuous low-thrust ion engines. All these approaches require specific new trajectory designs. There is increasing effort in applying non-linear, non-Keplerian orbits. If you have continuous low-level thrust, you have a non-linear problem in infinite dimensions.

Off into Deep Space

Solar Sails might be a good way to accelerate to a fraction of the speed of light, the visit nearby star seasons within our lifetimes. Of course, the further the mission gets away from the Sun, the less propulsion it sees. The trick is to accelerate as much as possible. There is little or no drag – but the problem is how to slow down at the target star location – just turn around and use the sail as a break. Another option is to use large solar-powered lasers to keep the sailcraft going, augmented by a large, orbiting Fresnel lens. We don't know how to build all of this, but we're learning.

In the 1970, Robert Forward described a laser installation to provide a power source for distant light-sailers in his science fiction book, Rocheworld. You want to go really fast? How about using a solar-sailing craft at a supernova?

Now it starts getting a little complicated. We want to touch on e-sails, and m-sails, better equipped for interstellar journeys.

An electric sail uses the dynamic pressure of the solar

wind as a source of thrust. But, it is constructed small wires to from an electric field, in effect, a virtual sail. The idea was suggested in 2006 in Finland. It is very different than the traditional solar sail which uses the momentum of electrons impinging on the sail. The E-sail uses an electric field to gain momentum by interfacing with solar wind ions. The pressure is only about 1% of the traditional solar sail. However, it can continue to accelerate at greater distances from the Sun. An E-sail equipped nanosat, *Aalto-2*, was launched in 2017, and tested the concept for de-orbiting in 2019. It worked fine.

A magnetic sail uses a static magnetic field to deflect charged particles in a plasma, gathering momentum from the process. As opposed to a standard lightsail, these do use charged particles for momentum. They could brake against any planet with a magnetosphere.

At the Earth's distance from the Sun, the solar wind consists several million protons and electrons per cubic meter, flowing at 4-600 km/sec. The Magsail uses a magnetic field to deflect some of these, thus gaining momentum. To make the mag-sail practical, we would need a high temperature superconductor. So, Mag-sails use the plasma, and solar sails use the photos emitted from the Sun. Both methods have a generated thrust that fall off as the square of the distance from the Sun. Mag-sails can also tack.

Breakthrough Starshot is a very well funded project, to develop a fleet of 1,000 nanocraft with lightsails. These would be propelled out of our solar system by a ground based multi-gigawatt laser, and head to Alpha Centauri, achieving 20% of the speed of light. Travel time would be 20 years.

My suggestion is to have one large "mothership" with 1,000 Cubesats, going to the edge of our solar system on a standard lightsail, and by then the laser technology would be worked out. At Alpha Centauri, it is feasible to use that star's solar wind to break (I was going to say "aerobrake," but its actually astro-brake. At this point the swarm could be released.)

We could envision a large laser facility, using solar power to direct a beam of energy towards a craft's solar sail. This might be a way to get out of our solar system, and head out to other ones. There is still the problem of braking.

Next-Gen solar sails

Look for new sail meta-materials that alter the angle of incoming photons, effectively, a diffraction grating. The advantage is, a diffraction sail does not heat up in the interaction with the photons. Sails will get thinner, stronger, and lighter. Most proposed sails are square kilometers of microscopically thin material, with reflective metallic surfaces.

Diffraction sails don't use metallic coatings, but rather meta-materials for textiles that are used to change the angle of incoming photons, forming basically, a diffraction grating. One advantage is, the sail does not heat up, as does a simply reflective one.

Another idea that has been kicked around is that of

optical. This would use a wing-shaped refractive sail which can fly in a uniform stream of light. More like a solar glider than a solar sail,

A mission was considered, to address the problem that the solar poles have never been imaged.

Don't just send one

Here we discuss aggregations of Cubesats. They may interact with each other, or not. Some of the architectures include trains, Constellations, and clusters. Groups of Cubesat explorers can Conduct radio occultation experiments to better categorize the distribution of particles. They can also perform synchronized simultaneous observation from multiple observation points.

They can map magnetic fields and charged particle environments. They will be able to examine phenomena of opportunity, as they arise. They can respond to targets of opportunity, such as the observed plunge of comet Shumaker-Levy-3 into Jupiter's atmosphere by the Galileo spacecraft.

Enabling Technologies

This section will discuss some of the enabling technologies that will be used in the Deep Space Cubesat Missions.

Overall Architecture

In this concept, the Cubesats are the primary payload.

The Mothership can be thought of as a very large Cubesat. The architecture is kept as close as possible.

Use of a common hardware bus and software architecture for all swarm members, to the greatest extent possible, is a goal. Only the sensor sets will be unique. A Cubesat model for the hardware, and NASA GSFC's Core Flight Software is baselined. A standard linux software operating environment and database will be used.

Each member of the swarm will be aware of other swarm members in close proximity. This will be facilitated by having the Mothership as the center of the coordinate system. It will determine its position by celestial navigation. The Cubesats will use the mothership their reference. The mothership will maintain, as part of its onboard database, the location of swarm members. It will also monitor for pending collisions and warn the participants. There will be rules concerning how close swarm members can get to each other, a virtual zone of exclusion. All Earth-based interaction with the swarm will be through the Mothership. Due to varying communication delays, tele-operation of the swarm from Earth is not feasible. The Swarm could be on the opposite side of the Sun from the Earth for extended periods. This is addressed by building autonomy into the system, and a large amount of non-volatile storage will be included for science data.

Each swarm member will be equipped with one or more cameras, not only for target investigation, but also for observing the position and relative motions of other swarm members.

Using standard linux clustering software (Beowulf), the Mothership and undeployed swarm members will be able to form an ad-hoc cluster computer to process science data in-situ. Within the Mothership, a LAN-based Mesh network software will be used. The Mothership's main computer will be a Raspberry-Pi based compute-cluster.

Complexity in a system generally derives from two parameters, the number of units, and the number of interactions. A swarm of Cubesats is complex, compared to a single spacecraft. This is balanced by the relative simplicity of the individual units, their standardization, flexibility, and redundancy. Redundancy is at the Cubesat level.

Mothership

The mothership will be built with standard aerospace products An X-band transceiver or laser would be a candidate for the Earth link. Standard p-pod cubesat dispensers are baselined, but the affects of long storage of the CubeSats in the dispensers must be considered. The effects of cold welding during the multi-year transit needs to be studied. The problem is that the Cubesat may not deploy, due to cold-welding of it to the dispenser. Cold welding occurs with similar materials touching in a vacuum.

The mothership would follow a Cubesat architecture for its computation, data storage, and telecommunications. It can also host sensors. The Cubesats will be identical on the spacecraft buss side, hosting differing science instruments.

The Mothership provides cloud services to the swarm. It is a store-and-forward node, and the communications relay to Earth. It provides Swarm control, monitoring, task assignment, and is responsible for Science and engineering data storage.

Onboard databases

Each member of the Swarm is self-documenting. It carries a copy of its Electronic Data Sheet (EDS) description, which can be updated. This defines the system architecture and capabilities, and has both fixed (as-built) and variable entries. The main computer in the Mothership has a copy of all of these, and can get updates by query. The Mothership also has parameters on each unit's state, such as electrical power remaining, temperature, position, etc. One value of the database is, if the Mothership needs a unit with a high resolution imager, it knows what unit that is, and whether it has been deployed or not. If it has been deployed, it will query the unit on its position and health status. Implementing the EDS in a true database has big advantages, since the position of the data item in the database also carries information. It also allows the use of off-the-shelf database tools. The individual Cubesats have a "light-weight" version of the database, while the Mothership has a more sophisticated one. All the schema's are the same. The advantage of a formal database is the structure it imposes on the data

There are two parts of the tables, representing static and dynamic data. Static Data represents the hardware and software configuration of the swarm unit. These values are not expected to change during the unit's operation. The Dynamic Data table represents the sensors each particular unit has. These values can change, and the last values will be kept. Cubesats will exchange two types of data through their communication channel: primary observational data, along with secondary metadata which includes position and localization information, along with timing information as a part of the EDS during the mission. This approach was prototyped in a previous project.

The Mothership is responsible for aggregating all of the Cubesats' housekeeping and science data, and transmitting it back to Earth. This is also facilitated by the structure imposed by the database. An Open Source version of an SQL database will be implemented. The EDS documents will be in XML, and the probable database is mySQL, which also has a *light* version.

Data compression can be implemented onboard the mothership , as well as preliminary data analysis for replanning.

Communications

NASA's Deep Space Network consists of three sites spaced around our planet. It supports deep space missions for NASA and other entities. It is managed by the Jet Propulsion Lab (JPL) in Pasadena, California. The nearest station to JPL is at Goldstone, in the desert to the east. Two other stations, in Spain and Australia are spaced about 120 degrees apart on the globe from Goldstone. The DSN started operations in the 1960's, with teletype communications with the Pasadena facility. The DSN is heavily oversubscribed, supporting numerous deep space missions.

Several approaches to communication with spacecraft at a large distance from Earth, and examining other planets have been defined. The Interplanetary Internet implements a Bundle Protocol to address large and variable delays. Normal IP traffic assumes a seamless, end-to-end, available data path, without worrying about the physical mechanism. The Bundle protocol addresses the cases of high probability of errors, and disconnections. This protocol was tested in communication with an Earth orbiting satellite in 2008.

As we get farther from Earth, the Cubesat's small antennas, and relatively low power, means we have to get clever with communications. There will be a limited bandwidth. This happened with the New Horizon's spacecraft at Pluto – It took more than 16 months to transmit all the data back. JPL's new approach will use laser communications.

Communications between planets in our solar system involves long distances, and significant delay. New protocols were needed to address the long delay times, and error sources.

A concept called the Interplanetary Internet uses a store-and-forward node in orbit around a planet (initially, Mars) that burst-transmits data back to Earth during available communications windows. At certain times, when the geometry is right, the Mars bound traffic might encounter significant interference. Mars surface craft communicate to Orbiters, which relay the transmissions to Earth. This allows for a lower wattage transmitter on the surface vehicle. Mars does not (yet) have the full infrastructure that is currently in place around the Earth – a network of navigation, weather, and communications satellites.

For satellites in near Earth orbit, protocols based on the cellular terrestrial network can be used, because the delays are small. In fact, the International Space Station is a node on the Internet. By the time you get to the moon, it takes about a second and a quarter for electromagnetic energy to traverse the distance. Delay tolerant protocols developed for mobile terrestrial communication were used, but break down in very long delay situations.

We have a good communications model and a lot of experience in Internet communications. One of the first implementations for space used a File Transfer Protocol (FPP) running over the CCSDS space communications protocol in 1996.

The formalized Interplanetary Internet evolved from a study at JPL, lead by Internet pioneer Vint Cerf, and Adrian Hook, from the CCSDS group. The concepts evolved to address very long delay and variable delay in communications links. For example, the Earth to Mars delay varies depending on where each planet is located in its orbit around the Sun.

The Interplanetary Internet implements a Bundle Protocol to address large and variable delays. Normal IP traffic assumes a seamless, end-to-end, available data path, without worrying about the physical mechanism. The Bundle protocol addresses the cases of high probability of errors, and disconnections. This protocol was tested in communication with an Earth orbiting satellite in 2008.

NASA's Goddard Space Fight Center in Greenbelt, Maryland, has been the hub of the space data network since the beginning. In the Apollo era, a world-wide system of ground stations providing continuous coverage was not yet in existence. NASA supplemented their ground stations with a series of tracking ships, to fill in coverage gaps. All data came to the basement of the Operations Building, 14, at Greenbelt. It was then routed upstairs to satellite control centers, or to Houston or Marshall for Manned flights. For its interplanetary missions, JPL maintained the Deep Space Network, a set of three very large antennas spaced around the world. During launch and near Earth operations, these were supplemented by NASA's world-wide set of tracking stations for Earth orbiting satellites.

ESA has published a road map of Interplanetary Missions. There are addressing Near Earth Objects, Venus, and Mars, in addition to lunar missions. They have the concept of a mothercraft for transportation and data relay to Earth.

Rad-Hard Software

This is a concept that implements routines that check and self-check, report, and attempt to re-mediate radiation damage. It is an outgrowth of the testing and self-testing of a computers' functionality, with focus on detection of radiation induced damage. We know, for example, that one of the tell-tales for radiation damage is increasing current draw. At the same time, we monitor other activities and parameters in the system. This partially addresses the problem of operating with non-radiation hardened hardware in a high radiation environment. The baseline Raspberry-Pi has been radiation tested to 150 kRad, and was operational at that point.

From formal testing results, and key engineering tools, we define likely failure modes, and possible remediation's. Besides self-test, we will have cross-checking of systems. Not everything can be tested by the software, without some additional hardware. First, we use engineering analysis that will help us define the possible hardware and software failure cases, and then define actions and remediation. This is a software FMEA.

failure modes and effects analysis. None of this is new, and the approach is to collect together best practices in the software testing area, develop a library of RHS routines, and get operational experience. Another advantage of the software approach is that we can change it after launch, as more is learned, and conditions change.

Rad Hard software runs in the background on the flight computer, and checks for the signs of pending failure from any known cause. The biggest indicator for radiation damage is an increase in current draw. The mothership cpu cluster monitors and trends current draw across the swarm, and take critical action such as a reboot if it deems necessary. The Rad Hard software will keep tabs on memory by conducting continuous CRC (cyclic redundancy checks). One approach to mitigating damage to semiconductor memory is "scrubbing," where we read and write back each memory locations (being careful not to interfere with ongoing operations). This will be done by a background task that is the lowest priority in the system. Watchdog timers are also useful in getting out of a situation such as a Priority Inversion, or just a radiation-induced bit flip. There will be a pre-defined safe mode for the computer as well. Key state data from just before the fault will be stored. Unused portions of memory can be filled with bit patterns that can be monitored for changes. We must be certain that all of the unused interrupt vectors point to a safe area in the code, so this will be reloaded periodically.

Functions within the RHS include current monitoring as a tell-tale of radiation damage, self-diagnosis suite, spurious interrupt test, memory test(s), checksums over code, data corruption testing, memory scrub, I/O functionality test, peripherals test, stack overflow monitoring, and a watchdog timer. A complete failure modes and effects analysis will be conducted over the flight computer and associated sensors and mechanisms, and this will be used to scope the RHS. The systems will keep and report trending data on the flight electronics. In most cases, the only remediation is a reboot. However, since the units will have identical configurations, the data will be useful to be able to predict pending failures, and to possibly avoid and correct them. This will be used on the Mothership's and on the Cubesat's RaspberryPi-based flight computers. This provides a distributed fault detection and mitigation system, with learning.

We can also choose to implement a small, rad-hard recovery computer, such as a Rad-hard Arduino Microcontroller, which uses FRAM, and is fairly immune to radiation. The recovery computer would receive heart-beat signals from the cluster members onboard the mothership, and take recovery efforts if they are interrupted. A similar scheme could be used onboard the Cubesats, with little impact on size, weight, power, and cost. This would primarily be used to mitigate latchup. At the distances we are interested in, the galactic cosmic rays would be problematic.

Autonomous operations

Where we are sending these Cubesats, they are on their own. Communications time precludes troubleshooting and updates from Earth. The explorer's are operating autonomously. The approach is the mothership,is in charge.

Making the Mothership in charge means it will aggregate all of the data from the other units, and send it back to Earth. It will have a large x-band of laser communication link, and will be sending continuous data. It will store the data onboard until the receipt has been acknowledged. This can span multiple years.

When NASA's *New Horizons* spacecraft imaged Pluto, it took 14 months for all the data to be received at Earth. Over the immense distances, the one-way light time is a factor, but so also is the available power for the transmitter.

Fly the Control Center

The Mothership is the navigation reference point for the Cubesats. It obtains its position with respect to Earth from observation, and ground tracking. There will be times when the Earth is not visible form the Mothership's position, so it will use extrapolation and local observation. During these periods of occultation, and also periods of long one-way light times, the Mothership assumes local responsibility for the Health and Safety of the Swarm members, and operations of the Swarm. For this, we will implement Control Center functionality within the Mothership. This will take the form of Ball Brother's COSMOS software. This product addresses traditional system test, integration, and flight needs. An additional software module is needed, essentially a virtual Control System Operator. Using defined rules, the Mothership will make decisions concerning the Swarm

Members, to the best of its current knowledge. All of this will be documented and downloaded to the Earth-based control center when communications is re-established. An AI capability will be added to Cosmos, in the form of a virtual flight controller agent. Besides the housekeeping functions, we will implement onboard science planning, responsive to on-site conditions, and targets of opportunity.

The Mothership's primary responsibility is continuance of the Mission. To a degree, the Cubesats are considered expendable. During communications black-outs, observations will continue, and the Mothership will dispense explorers according to pre-defined rules, and based on it's best on-scene judgment. It will also continue to collect observation science data, and engineering data related to health and performance across the swarm members.

Mission goals may change based on observations. This is where the agile collection of explorers can make a big difference

The Mothership's primary responsibility is Continuance of the Mission. To a degree, the Cubesats are considered expendable. During communications black-outs, observations will continue, and the Mothership will dispense explorers according to pre-defined rules, and based on it's best on-scene judgment. It will also continue to collect observational science data, and engineering data related to health and performance across the swarm members.

Swarm Architecture

Each member of the swarm will be aware visually of other swarm members in close proximity. This will be facilitated by having the Mothership as the center of the coordinate system. It will determine its position by celestial navigation. The Cubesats will have a similar capability. The mothership will maintain, as part of its onboard database, the location of all other members. It will also monitor for pending collisions and warn the participants. There will be rules concerning how close swarm members can get to each other, a virtual zone of exclusion. All Earth-based interaction with the swarm will be through the Mothership.

This section describes a different approach: collections of smaller co-operating systems that can combine their efforts and work as ad-hoc teams on problems of interest. Cubesats can be organized in Swarms.

This is based on the collective or parallel behavior of homogeneous systems. This covers collective behavior, modeled on biological systems. Examples in nature include migrating birds, schooling fish, and herding sheep. A collective behavior emerges from interactions between members of the swarm, and the environment. The resources of the swarm are organized dynamically. A swarm of Cubesats with differing capabilities can be used, combining into Teams of Convenience to address situations and issues discovered in situ.

Biological swarms, such as ants, achieve success by

division of labor throughout the swarm, collaboration, and sheer numbers. They have redundancy, as any individual can do any task assigned to the swarm. The individual units are highly autonomous, but are dependent on other members for their needs. They achieve success with a simple neural architecture and primitive communications.

In Swarm robotics, the key issues are communication between units, and cooperative behavior. The capability of individual units nodes not much matter; it is the strength in numbers. Ants and other social insects such as termites, wasps, and bees, are models for robot swarm behavior. Self-organizing behavior emerges from decentralized systems that interact with members of the group, and the environment. Swarm intelligence is an emerging field, and swarm robotics is in its infancy. Co-operative behavior, enabled by software and intra-unit communications has been demonstrated.

So, what do we use for the Swarm software. That is also NASA Open Source, and off the shelf. NASA did a lot of work in the 1970's and 1980's on ANTS – the Autonomous Nano-Technology Swarm. The Basic concepts were defined, but the implementation was not yet ready. Now, Cubesat-type architectures can address that. The overall concept is many small units working together, self-configuring, self-healing, making in-situ decisions. No direct human control, not per-programmed, but agile, making decision on the spot. A reference mission to the Asteroid Belt in 2020 was defined. Well, here we are, where's the mission? (This is partially addressed in the authors book, "A Cubesat Swarm Approach for Exploration of the Asteroid Belt, originally a student project that got to a high TRL). Another feature the swarm exhibits is teaming, social interaction between individual units. ANTS was hard to implement with the hardware and software available in the 1980's. It fits well into a Cubesat implementation.

Avionics Suite

Both the Mothership and the Cubesats will baseline the GSFC PiSat software and hardware architecture for the flight computers. The Cubesats will use a single unit, and the Mothership will have a 16-unit compute cluster. Non-deployed Cubesats in the Mothership will be able to participate in the clustering, using the Mothership's internal networking infrastructure. The Mothership will be able to power up and attach selected additional units for particularly computational-intensive tasks.

The Raspberry Pi-3 is a very capable processor. An earlier model was tested to operate to 150 k Rad Total Ionizing Dose, with only the loss of several unused I/O interfaces. The major source of radiation at the destination will not be trapped particles, but rather ionizing cosmic rays of galactic origin. These are energetic, but sparse. The cluster computer will be enclosed in the nose of the mothership, with shielding.

A Raspberry Pi-3, requires 3.26 watts of power. It is quad-core, operating at 1.4 GHz. It is a 64-bit machine, with 1 gigabyte of ram, and can achieve 2451 MIPS. It has a dedicated Graphics Processing Unit-based video pipeline that can handle 2D DSP, supported by the Open-GL software library.

Compute cluster of convenience

Using a variation of the Beowulf clustering software and the communications infrastructure of the Mothership, the Cubesats awaiting deployment can be linked into the Mothership's Compute Cluster of Convenience configuration. Each compute node will have the Beowulf software pre-loaded as part of its Linux operating system.

Beowulf was developed at GSFC to provide a low cost solution to linking commodity pc's into a supercomputer. The approach has been applied to clusters of small architectures such as the Raspberrry-Pi, that serve as flight computers for Cubesats. Several 64-node Pi clusters have been demonstrated in the Earth environment.

The Beowulf cluster is ideal for sorting and classifying data; an example application for this is the Probabilistic Neural Network (PNN). This algorithm has been used to search for patterns in remotely sensed data. It is computationally intensive, but scales well across compute clusters. It was developed by the Adaptive Scientific Data Processing (ASDP) group at NASA/GSFC. The program is available in Java source code.

The first Beowulf cluster to be flown in space was built

from twenty 206-MHz StrongARM (SA1110) processors, and flew on the X-Sat, Singapore's first satellite. The performance was 4,000 MIPS. The cluster drew 25 watts. The satellite was a 100 kg, 80 CM cube. The cluster was used because the satellite collected large amounts of image data (80 GB per day), most of which was not relevant to the mission. An onboard classification algorithm selected which images would be downloaded. For example, cloudy images were discarded, since only land images of Singapore were of interest.

In a cluster, there is always a trade-off of computation, communication, and power draw. This will be monitored and adjusted by the cluster itself, in real-time.

In Situ data processing

Using standard linux clustering software (Beowulf), the Mothership will be able to process and store science data in-situ. Within the Mothership, the use of LAN-based Mesh network is feasible The Mothership's main computer will host Raspberry-Pi based cluster, probably 64 nodes. This will be loaded with the NASA-developed Beowulf software, running parallel algorithms to analyze the science data. It will have certain elements that will be looking for trends and outliers (SP).

Glossary of Terms

U – one unit for a Cubesat, 10 x 10 x 10 cm.

3U – three units for a Cubesat

6U – 6 units in size, where 1u is defined by dimensions and weight.

802.11 – a radio frequency wireless data communications standard.

AACS – (JPL) Attitude and articulation control system.

ACE – attitude control electronics

AGC – Automated guidance and control.

AI – Artificial Intelligence

AIAA – American Institute of Aeronautics and Astronautics.

AIST – NASA GSFC Advanced Information System Technology .

ANTS – autonomous nano technology swarm

Antares – Space launch vehicle, compatible with Cubesats, by   Orbital/ATK (U.S.)

AP – application programs.

Apm – antenna pointing mechanism

Arduino – a small, inexpensive microcontroller architecture.

ASIC – application specific integrated circuit

ASIN – Amazon Standard Inventory Number.

Asteroid - minor planets, orbiting the Sun.

async – non synchronized

ATP – authority to proceed

AU – astronomical unit. Roughly 149.6 million kilometers, the mean distance between Earth and Sun.

BAE – British Aerospace.

Baud – symbol rate; may or may not be the same as bit rate.

Beowolf – a cluster of commodity computers; multiprocessor, using Linux.

BIST – built-in self test.

Bit – binary variable, value of 1 or 0.

Bow shock- Where the solar wind begins to interact with a planet's magnetosphere.

Centaur – a minor planet in an unstable orbit, behaving like an asteroid or comet.

Comet – icy body orbiting the Sun inn a very eccentric orbit.

BP - bundle protocol, for dealing with errors and disconnects.

Bus – an electrical connection between 2 or more units; the engineering part of the spacecraft.

CalPoly – California Polytechnic State University,. San Luis Obispo, CA.

CAN - controller area network bus.

CCSDS – Consultive Committee on Space Data Systems.

CDR – critical design review

C&DH – Command and Data Handling

CDFP - CCSDS File Delivery Protocol

Centaur – a minor planet in an unstable orbit, behaving like an asteroid or comet.

Comet – icy body orbiting the Sun inn a very eccentric orbit.

cFE – Core Flight Executive – NASA GSFC reusable flight software.

CFS – Core Flight System – NASA GSFC reusable

flight software.

CME – Coronal Mass Ejection. Solar storm. .

CogE – cognizant engineer for a particular discipline; go-to guy; specialist.

Comet – icy body orbiting the Sun inn a very eccentric orbit.

Constellation – a grouping of satellites.

COTS – commercial, off the shelf

CPU – central processing unit

CRC – cyclic redundancy code – error detection and correction mechanism.

Cubesat – small inexpensive satellite for colleges, high schools, and individuals.

DARPA – (U. S.) Defense advanced research projects agency.

Dnepr – Russian space launch system compatible with Cubesats.

DOD – (U. S.) Department of Defense.

DOE – (U. S.) Department of Energy.

DOF – degrees of freedom.

Downlink – from space to earth.

DSN – Deep Space Network

DSP – digital signal processing/processor.

DTE – Direct-to-Earth

DTN – delay tolerant network; disruption tolerant network.

ECC – error correcting code

Ecliptic – apparent path of the Sun throughout the year.

EDAC – error detection and correction.

EDS – Electronic Data Sheets

EGSE – electrical ground support equipment

ELV – expendable launch vehicle.

Embedded system – a computer systems with limited human interfaces and performing specific tasks. Usually part of a   larger system.

EMC – electromagnetic compatibility.

EMI – electromagnetic interference.

EOL – end of life.

Ephemeris – orbital position data.

EPS – electrical power subsystem.

ESA – European Space Agency

ESRO – European Space Research Organization

ESTO – NASA/GSFC – Earth Science Technology Office.

ev – electron volt, unit of energy

EXPRESS racks – on the ISS, EXpedite the PRocessing of Experiments for Space Station Racks

Falcon – launch vehicle from SpaceX.

FCC – (U.S.) Federal Communications Commission.

FDC – fault detection and correction.

Flag – a binary state variable.

Flash – non-volatile memory

Flatsat – prototyping and test setup, laid out on a bench for easy access.

FlightLinux – NASA Research Program for Open Source code in space.

Floating point – computer numeric format for real numbers; has significant digits and an exponent.

FPGA – field programmable gate array.

FPU – floating point unit, an ALU for floating point numbers.

Full duplex – communication in both directions

simultaneously.

FRAM – ferromagnetic RAM; a non-volatile memory technology

FRR – Flight Readiness Review

FSW – flight software.

FTP – file transfer protocol

Gbyte – 10^9 bytes.

GeV – Giga (10^9) electron volts.

Giga - 10^9

GNC – guidance, navigation, and control.

GPIO – general purpose I/O.

GPL – gnu public license used for free software; referred to as the "copyleft."

GPU – graphics processing unit. ALU for graphics data.

GSFC – Goddard Space Flight Center, Greenbelt, MD.

Gyro – (gyroscope) a sensor to measure rotation.

Handshake – co-ordination mechanism.

HDL – hardware description language

Hi-rel – high reliability

IARU – International Amateur Radio Union

IAU – International Astronomical Union

ICD – interface control document.

IC&DH – Instrument Command & Data Handling.

Ice giant – A large icy/liquid planet, consisting of elements heavier   than hydrogen and helium.

IEEE – Institute of Electrical and Electronic engineers

IMU – inertial measurement unit.

Integer – the natural numbers, zero, and the negatives of the natural numbers.

Interrupt – an asynchronous event to signal a need for attention (example: the phone rings).

IP – intellectual property; Internet protocol.

IP core – IP describing a chip design that can be licensed to be used in an FPGA or ASIC.

IP-in-Space – Internet Protocol in Space.

IR – infrared, 1-400 terahertz. Perceived as heat.

ISA – instruction set architecture, the software description of the computer.

ISBN – International Standard Book Number.

ISO – International Standards Organization.

ISR – interrupt service routine, a subroutine that handles a particular interrupt event.

ISS – International Space Station

I&T – integration & test

ITAR – International Trafficking in Arms Regulations (US Dept. of State)

ITU – International Telecommunications Union

IV&V – Independent validation and verification.

JEM – Japanese Experiment Module, on the ISS.

JHU – Johns Hopkins University.

JOI – Jovian orbit insertion,

Jovian – pertaining to Jupiter.

JPL – Jet Propulsion Laboratory

JSC – Johnson Space Center, Houston, Texas.

JTAG – Joint Test Action Group; industry group that lead to IEEE 1149.1, Standard Test Access Port and Boundary-Scan Architecture.

JWST – James Webb Space Telescope – follow on to Hubble.

KBO – Kuiper belt object.

Kbps – kilo (10^3) bits per second.

KBO – Kuiper Belt Object

Kg – kilogram.

kHz – kilo (10^3) hertz

Kuiper Belt – beyond Neptune, a ring of small icy asteroids and minor planets.

Ku band – 12-18 Ghz radio

Lagrange (L) point – a null point in the gravity field in the 3-body program.

    L1 - the Lagrange point between the 2 bodies.

    L2 – the Lagrange point behind the smaller body.

    L3 – the Lagrange point behind the larger body.

    L4- the leading Lagrange in an orbit.

    L5 – the trailing Lagrange point in an orbit.

Lan – local area network, wired or wireless.

LaRC – (NASA) Langley Research Center.

Latchup – condition in which a semiconductor device is stuck in one state.

Lbf – pounds-force (0.7 newton-meter)

LEO – low Earth orbit.

Let- Linear Energy Transfer

LGM – little green men.

Lidar – optical radar.

Linux – open source operating system

LRR – launch readiness review

LRU – least recently used; an algorithm for item replacement in a cache.

LSB – least significant bit or byte.

LSP – (NASA) launch services program, or launch provider

LUT – look up table.

Magnetosphere – a space surrounding a planet or moon that is affected by the primary's magentic field.

Magentopause – abrupt boundary between a magnetosphere and the solar wind.

Magnetotail – magnetosphere extends away from the planet and the Sun.

Master-slave – control process with one element in charge. Master status may be exchanged among elements.

Mbps – mega (10^6) bits per second.

Mbyte – one million (10^6 or 2^{20}) bytes.

Memory leak – when a program uses memory resources but does not return them, leading to a lack of available memory.

Memory scrubbing – detecting and correcting bit errors.

MEMS – Micro Electronic Mechanical System.

MESI – modified, exclusive, shared, invalid state of a cache coherency protocol.

MEV – million electron volts.

MHz – one million (10^6) Hertz

Microcontroller – monolithic cpu + memory + I/O.

Microkernel – operating system which is not monolithic, functions execute in user space.

Microprocessor – monolithic cpu.

Microsat – satellite with a mass between 10 and 100 kg.

Microsecond – 10-6 second.

MJS-77 circa 1977 mission to Mars, Jupiter, Saturn. Name changed to Voyager.

MLI – multi-layer insulation.

MPA – multiple payload adapter for deploying multiple p-pod's

MPE – Maximum predicted environments.

Mph – miles per hour

mram – magnetorestrictive random access memory.

mSec – Millisecond; (10^{-3}) second.

MIPS – millions of instructions per second.

MMU – memory management unit; manned maneuvering unit.

MSB – most significant bit or byte.

Multiplex – combining signals on a communication channel by sampling.

Multicore – multiple processing cores on one substrate or chip; need not be identical.

Mutex – a software mechanism to provide mutual exclusion between tasks.

Nano – 10^{-9}

NanoRacks – a company providing a facility onboard the ISS to   support Cubesats

nanoSat – small satellite with a mass between 1 and 10 kg.

NASA - National Aeronautics and Space Administration.

NDA – non-disclosure agreement; legal agreement protecting IP.

NEA – near Earth asteroid

NEC – near Earth Comet

NEN – (NASA's) Near Earth Network

NEO – near Earth object.

Nibble – 4 bits, ½ byte.

NIST – (U.S.) National Institute of Standards and Technology, previously, National Bureau of Standards.

NMI – non-maskable interrupt; cannot be ignored by the software.

NOAA – (U.S.) National Oceanographic and

Atmospheric Administration.

Normalized number – in the proper format for floating point representation.

NRCSD - NanoRack CubeSat Deployer

NRE – non-recurring engineering; one-time costs for a project.

NSF – (U.S.) National Science Foundation.

NSR – non-space rated.

NTIA (U.S.) National Telecommunications and Information Administration

NUMA – non-uniform memory access for multiprocessors; local and global memory access protocol.

NVM – non-volatile memory.

NWS – (U.S.) National Weather Service

Nyquist rate – in communications, the minimum sampling rate, equal to twice the highest frequency in the signal.

OBC – on board computer

OBD – On-Board diagnostics.

OBP – On Board Processor

Off-the-shelf – commercially available; not custom.

One-way light time – a measure of distance, in terms of how long it would take light to travel the distance.

Orbit – the path of one body around another, that are linked by gravity.

OpAmp – operational amplifier; linear gain and isolation stage.

OpCode – encoded computer instruction.

Open source – methodology for hardware or software

development with free distribution and access.

Operating system – software that controls the allocation of resources in a computer.

OSAL – operating system abstraction layer.

OSI – Open systems interconnect model for networking, from ISO.

Overflow - the result of an arithmetic operation exceeds the capacity of the destination.

Packet – a small container; a block of data on a network.

Paging – memory management technique using fixed size memory blocks.

Paradigm – a pattern or model

Paradigm shift – a change from one paradigm to another. Disruptive or evolutionary.

Parallel – multiple operations or communication proceeding simultaneously.

Parity – a simple error detecting mechanism involving an extra check bit in the word.

PC-104 – standard for a board (90 x 96 mm), and a bus for embedded use.

PCB – printed circuit board.

pci – personal computer interface (bus).

PCM – pulse code modulation.

PCSI – NASA's Planetary Cubesat Science Institute

PDCO – NASA's Planetary Defense Coordination Office

PDR – preliminary design review

Perhelion – in an orbit, the closest point to the Sun.

Peta - 10^{15} or 2^{50}

PHO – potentially hazardous object

Phonesat – small satellite using a cell phone for onboard control and computation.

Picosat – small satellite with a mass between 0.1 and 1
    kg.
Piezo – production of electricity by mechanical stress.
Pinout – mapping of signals to I/O pins of a device.
Pipeline – operations in serial, assembly-line fashion.
PiSat – a Cubesat architecture developed at NASA-
GSFC, based   on the Raspberry Pi architecture.
Pixel – picture element; smallest addressable element on
    a display or a sensor.
PLL – phase locked loop.
PocketQube – smaller than a Cubesat; 5 cm cubed, a
    mass of no   more than 180 grams, and uses COTS
    components.
Poc – point of contact
POSIX – IEEE standard operating system.
PPF – payload processing facility
PPL – preferred parts list (NASA).
P-POD – Cubesat launch dispenser, Poly-Picosatellite
    Orbital Deployer
Psia – pounds per square inch, absolute.
PSP – Platform Support Package.
Rad – unit of radiation exposure
Rad750 – A radiation hardened IBM PowerPC cpu.
Ram – random access memory.
RBF – remove before flight.
Real-time – system that responds to events in a
    predictable, bounded time.
Reset – signal and process that returns the hardware to a
    known, defined state.
RF – radio frequency
RFC – request for comment

RHS – rad hard software

Ring system – a disk of solid material around a planet.

RTC – real time clock.

RTG – Radioisotope Thermal Generator – electrical power plant

RTOS – real time operating system.

SDR – software defined radio

SDRAM – synchronous dynamic random access memory.

Segmentation – dividing a network or memory into sections.

Semiconductor – material with electrical characteristics between conductors and insulators; basis of current technology for processor, memory, and I/O devices, as well as sensors.

Semaphore – a binary signaling element among processes.

SD – secure digital (non-volatile memory).

SDVF – Software Development and Validation Facility.

Sensor – a device that converts a physical observable quantity or event to a signal.

Serial – bit by bit.

SEU – single event upset (radiation induced error).

Servo – a control device with feedback.

Six-pack – a six U Cubesat, 10 x 20 x 30 cm.

SMP – symmetric multiprocessing.

Snoop – monitor packets in a network, or data in a cache.

SN – (NASA's) Space Network

SOA – safe operating area; also, state of the art.

SOC – system on a chip; also state-of-charge.

Socket – an end-point in communication across a

network

Soft core – a hardware description language description of a cpu core.

Software – set of instructions and data to tell a computer what to do.

SOI – Saturn Orbit insertion

Solar flare – a sudden rapid emission of electrons, ions, and atoms from a star.

Solar System – A star and its associated planets and such.

Solar wind – stream of charged particles emitted from a star's upper atmosphere.

SMP – symmetric multiprocessing.

Snoop – monitor packets in a network, or data in a cache.

Spacewire – high speed (160 Mbps) link.

SPI - Serial Peripheral Interface - a synchronous serial communication interface.

SRAM – static random access memory.

STAR – self test and repair.

State machine – model of sequential processes.

STOL – system test oriented language, a scripting language for testing systems.

Strawman – an early concept or prototype, to be refined.

SWAP – size, weight, power

T&I – test and integration.

Terrabyte – 10^{12} bytes.

SAA – South Atlantic anomaly. High radiation zone in Earth's atmosphere.

SEB – single event burnout.

SEU – single event upset.

SEL – single event latchup.

Soc – state of charge; system on a chip.

Soft core – hardware description description language model of a   logic core.

SOI – silicon on insulator

SoS – silicon on sapphire – an inherently radiation-hard technology

spi – serial peripheral interface

SpaceCube – an advanced FPGA-based flight computer.

SpaceWire – networking and interconnect standard.

SRAM – static random access memory.

Stack – first in, last out data structure. Can be hardware or software.

Stack pointer – a reference pointer to the top of the stack.

State machine – model of sequential processes.

SWD – serial wire debug.

Synchronous – using the same clock to coordinate operations.

System – a collection of interacting elements and relationships with a specific behavior.

System of Systems – a complex collection of systems with pooled resources.

Suitsat – old Russian spacesuit, instrumented with an 8-bit micro, and launched from the ISS.

Swarm – a collection of satellites that can operate cooperatively.

sync – synchronize, synchronized.

TCP/IP – Transmission Control Protocol/Internet protocol.

TDRSS – Tracking and Data Relay Satellite System, Earth orbit.

Tera - 10^{12} or 2^{40}

Test-and-set – coordination mechanism for multiple

processes that allows reading to a location and writing it in a non-interruptible manner.

Thread – smallest independent set of instructions managed by a multiprocessing operating system.

TID – total ionizing dose.

Tidal lock – where the same side of a object always faces the primary it is orbiting.

TMR – triple modular redundancy.

TNO – Trans-Neptunian objects.

Toolchain – set of software tools for development.

Transceiver – receiver and transmitter in one box.

Transducer – a device that converts one form of energy to another.

Train – a series of satellites in the same or similar orbits, providing sequential observations.

TRAP – exception or fault handling mechanism in a computer; an operating system component.

Triplicate – using three copies (of hardware, software, messaging, power supplies, etc.). for redundancy and error control.

TRL – technology readiness level

Trojan - minor planet that shares an orbit with one of the larger planets.

Truncate – discard. cutoff, make shorter.

TT&C – tracking, telemetry, and command.

UDP – User datagram protocol; part of the Internet Protocol.

Underflow – the result of an arithmetic operation is smaller than the smallest representable number.

Uplink – from ground to space.

USAF – United States Air Force.

USB – universal serial bus.

VDC – volts, direct current.

Vector – single dimensional array of values.

VHDL – very high level design language.

Virtualization – creating a virtual resource from available physical resources.

Virus – malignant computer program.

WiFi – short range digital radio.

Watchdog – hardware/software function to sanity check the hardware, software, and process; applies corrective action if a fault is detected; fail-safe mechanism.

Wiki – the Hawaiian word for "quick." Refers to a collaborative content website.

Word – a collection of bits of any size; does not have to be a power of two.

X-band – 7 – 11 GHz.

Zombie-sat – a dead satellite, in orbit.

Zone of Exclusion – volume in which the presence of an object, personnel, or activities are prohibited

References

Aguilar, David A. *Space Encyclopedia: A Tour of Our Solar System and Beyond* (National Geographic Kids), 2013, ISBN-1426309481.

Alvarez, Jennifer L.; Rice, John R. Samson, Jr., Michael A. Koets. "Increasing the Capability of Cubesat-based Software-Defined Radio Applications," avail: ieeexplore.ieee.org/document/7500847/

Andrews, D. G. and Zubrin R. "NIAC Study of the Magnetic Sail,"avail: http://www.niac.usra.edu/files/library/meetings/fellows/nov99/320Zubrin.pdf

Asmar, Sami; Matousek, Steve "Mars Cube One (MarCO), First Planetary CubeSat Mission (presentation), 2014, JPL, avail: www.jpl.nasa.gov/cubesat/missions/marco.php

Asnuda, Paul A. "Open Courseware and STEM Initiatives in Career and Technical Education", avail, ir.library.illinoisstate.edu/cgi/viewcontent.cgi?article=1042&context=jste.

Baisamo, James M. et al "CubeSat technology adaption for in-situ characterization of NEOs," presentation, avail: NASA Technical Reports Server (NTRS), 2014, document id 20140004799.

Betancourt, Mark "CubeSats to the Moon (Mars and Saturn, too)", Air & Space Magazine, Sept 2014.

Budianu, A. et al.. "Inter-satellite links for Cubesats". In: IEEE Aerospace Conference Proceeding, 2013, pp. 1–10. avail: ieeexplore.ieee.org/document/6496947/

Challa, Obulapathi N., McNair, Janise "Distributed Data Storage on Cubesat Clusters," Advances in Computing 2013, (3) 3 pp.36-49. avail: http://article.sapub.org/10.5923.j.ac.20130303.02.html

Chujo, Toshihiro "Liquid Crystal Device with Reflective Microstructure for Attitude Control," 2018, avail: https://arc.aiaa.org/doi/abs/10.2514/1.A34165.

Clark, P. E.; et al *BEES for ANTS: Space Mission Application for the Autonomous NanoTechnology Swarm,* avail: https://arc.aiaa.org/doi/abs/10.2514/6.2004-6303.

Copernicus, Nicolaus *On the Revolutions of the Heavenly Spheres*, 1543, ASIN-B01MS8TGOV.

Dethloffy, Henry C.; Schorn, Ronald *Voyager's Grand Tour: To the Outer Planets and Beyond*, 2009, ISBN1568527152.

Engineering and Medicine, National Academies of Sciences, Division on Engineering and Physical Sciences, Space Studies Board, *Achieving Science with*

Cubesats: Thinking Inside the Box, National Academies Press, 2016, ISBN-978-0309442633.

Forward, R.L. (1984). "Roundtrip Interstellar Travel Using Laser-Pushed Lightsails".J Spacecraft. 21(2): 187–195, avail: https://arc.aiaa.org/doi/10.2514/3.8632.

Gilster, Paul "CubeSats: Deep Space Possibilities," Sept. 2015, avail: http://www.centauri-dreams.org/?p=34056.

Gilsdter, Paul "JAXA Sail to Jupiter's Trojan Asteroids," 2017, avail: https://www.centauri-dreams.org/2017/03/15/jaxa-sail-to-jupiters-trojan-asteroids/

Gilster, Paul, An Inflatable Sail to the Oort Cloud, 2008, avail: https://www.centauri-dreams.org/2008/11/12/an-inflatable-sail-to-the-oort-cloud/
Hall, John "maddog"; Gropp, William *Beowulf Cluster Computing with Linux*, 2003, ISBN-0262692929.

Hinchey, Michael G. ; Rash, James L.; Truszkowski, Walter E.; Rouff, Christopher A., Sterritt, Roy *Autonomous and Autonomic Swarms,* avail: https://ntrs.nasa.gov/search.jsp?R=20050210015-2017-12-20T20:19:24+00:00Z

Johnson, Les et al "Solar Sail Propulsion for Interplanetary Cubesats," NASA/MSFC.

Kreck Institute of Space Studies, *Small Satellites: A revolution in Space Science*, Final Report, July 2014. avail: kiss.caltech.edu/study/smallsat/KISS-SmallSat-FinalReport.pdf

Lappas, Vaios, et al CubeSail: A low cost Cubesat based solar sail demonstration mission, Advances in Space Research, 2001, 48.11, 1890-1901.

Macdonald, Malcolm "Advances in Solar Sailing,"2014, Springer Praxis, ISBN-978-3642349065.

Madni, Mohamed Atef; Raad, Raad; Tubbal, Faisal "Inter-Cubesat Communications: Routing Between Cubesat Swarms in a DTN Architecture," presentation, avail: https://iCubesat.org/papers/2015-2/2015-b-2-1.

Marchaj, C. A. *Sail Performance : Techniques to Maximize Sail Power*, 2002, ISBN-978-0071413107.

McLoughlin, Ian; Bretschneider, Timo; Ramesh, Bharath "First Beowulf Cluster in Space," Linux Journal, September 2005, Issue #137, article 8097.

Mori, Osamu, et al "System Designing of Solar Power Sail-craft for Jupiter Trojan Asteroid Exploration," avail: https://www.jstage.jst.go.jp/article/tastj/16/4/16_TJSAS-D-17-00070/_article/-char/ja/

NASA, "Hitchhiking Into the Solar System: Launching NASA's First Deep-Space Cubesats," avail:

www.nasa.gov/exploration.

NASA, *NASA's Great Observatories*, 2013, ASIN-B00DJUALH0.

Popescu, Otilia, "Power Budgets for Cubesat Radios to Support Ground Communications and Inter-Satellite Links" Department of Engineering Technology, Old Dominion University, Norfolk, avail: https://ieeexplore.ieee.org/document/7964683/

Sisson, Stephanie Roth *Star Stuff, Carl Sagan and the Mysteries of the Cosmos*, 2014, Roaring Brook Press, ISBN-978-159643-960-3.

Ross, Shane, "The Interplanetary Transport Network," American Scientist, Vol 94, May-June 2006.

Spangelo, et al "JPL's Advanced CubeSat Concepts for Interplanetary Science and Exploration Missions, Cubesat Workshop," 2015, California Institute of Technology, JPL. Avail:
https://digitalcommons.usu.edu/cgi/viewcontent.cgi?article=3313&context=smallsat

Staehle, Robert et al, "Interplanetary Cubesats: Opening the Solar System to a Broad Community at Lower Cost" Cubesat Workshop, 2011, Logan Utah.

Stakem, Patrick H.; Rezende, Aryadne; Ravazzi, Andre "Cubesat Swarm Communications," 2016.

Stakem, Patrick H.; Da Costa, Rodrigo Santos Valente; Rezende, Aryadne; Ravazzi, Andre "A Cubesat-based alternative for the Juno Mission to Jupiter, 2017, available from the author, pstakem1@jhu.edu.

Stakem, Patrick H. "Lunar and Planetary Cubesat Missions," March Volume 15, Polytech Revista de Tecnologia e Ciência,
avail:
http://www.polyteck.com.br/revista_online/ed_15.pdf

Stakem, Patrick H.; Martinez, José Carlos; Chandrasenan, Vishnu; Mittras, Yash; A Cubesat Swarm Approach for the Asteroid Belt," Presented to NASA Goddard Planetary CubeSats Symposium, August 16-17, 2018, NASA, GSFC, Greenbelt, MD.

Tonn, Shara, "How swarms of small satellites could revolutionize space exploration," Oct. 2016, Stanford Engineering, avail:
https://engineering.stanford.edu/news/

Truszkowski, Walt *Autonomous and Autonomic Systems: With Applications to NASA Intelligent Spacecraft Operations and Exploration Systems*, Springer; 1st Edition. edition, 2009, ISBN-1846282322.

Truszkowski, Walt, et al. *ANTS: Exploring the Solar System with an Autonomous Nanotechnology Swarm. J. Lunar and Planetary Science XXXIII (2002).*

Truszkowski, Walt "Prototype Fault Isolation Expert System for Spacecraft Control," N87-29136, avail: https://ntrs.nasa.gov/search.jsp?R=19870019703,

Violette, Daniel P. "Arduino/Raspberry Pi: Hobbyist Hardware and Radiation Total Dose Degradation, EEE Parts for Small Missions," GSFC, 2014, avail: https://ntrs.nasa.gov/search.jsp?R=20140017620.

Resources

- New Horizons Mission - https://www.nasa.gov/miss ion_pages/newhorizons/overview/index.html

- Cubesat Design Specification, Cubesat Program, California Polytechnic State University, avail.https://www.google.com/searchq=Cubesat+D esign+Specification&ie=utf-8&oe=utf-8

- https://www.planetary.org/explore/projects/lightsai l-solar-sailing/what-is-solar-sailing.html

- www.planetary.org

- NASA Systems Engineering Handbook, NASA SP-2007-6105.

- https://www.nasa.gov/mission_pages/

- http://astronomy.swin.edu.au/cosmos/C/Centaurs

- https://solarsystem.nasa.gov

- http://phys.org/news/2015-11-cubesats-deep-space.html

- NASA, Setting Sail for the Stars, 2000, avail: https://science.nasa.gov/science-news/science-at-nasa/2000/ast28jun_1m/

- "OPTIMIZED TRAJECTORIES TO THE NEAREST STARS USING LIGHTWEIGHT HIGH-VELOCITY PHOTON SAILS," avail: https://arxiv.org/pdf/1704.03871.pdf

- https://icubesat.files.wordpress.com/2019/05/a.1.1.201905271140-walker.pdf

- wikipedia, various.

If you enjoyed this book, you might also be interested in some of these.

Stakem, Patrick H. *16-bit Microprocessors, History and Architecture*, 2013 PRRB Publishing, ISBN-1520210922.

Stakem, Patrick H. *4- and 8-bit Microprocessors, Architecture and History*, 2013, PRRB Publishing, ISBN-152021572X,

Stakem, Patrick H. *Apollo's Computers,* 2014, PRRB Publishing, ISBN-1520215800.

Stakem, Patrick H. *The Architecture and Applications of the ARM Microprocessors,* 2013, PRRB Publishing, ISBN-1520215843.

Stakem, Patrick H. *Earth Rovers: for Exploration and Environmental Monitoring,* 2014, PRRB Publishing, ISBN-152021586X.

Stakem, Patrick H. *Embedded Computer Systems, Volume 1, Introduction and Architecture*, 2013, PRRB Publishing, ISBN-1520215959.

Stakem, Patrick H. *The History of Spacecraft Computers from the V-2 to the Space Station*, 2013, PRRB Publishing, ISBN-1520216181.

Stakem, Patrick H. *Floating Point Computation*, 2013,

PRRB Publishing, ISBN-152021619X.

Stakem, Patrick H. *Architecture of Massively Parallel Microprocessor Systems*, 2011, PRRB Publishing, ISBN-1520250061.

Stakem, Patrick H. *Multicore Computer Architecture,* 2014, PRRB Publishing, ISBN-1520241372.

Stakem, Patrick H. *Personal Robots*, 2014, PRRB Publishing, ISBN-1520216254.

Stakem, Patrick H. *RISC Microprocessors, History and Overview,* 2013, PRRB Publishing, ISBN-1520216289.

Stakem, Patrick H. *Robots and Telerobots in Space Application*s, 2011, PRRB Publishing, ISBN-1520210361.

Stakem, Patrick H. *The Saturn Rocket and the Pegasus Missions, 1965,* 2013, PRRB Publishing, ISBN-1520209916.

Stakem, Patrick H. *Visiting the NASA Centers, and Locations of Historic Rockets & Spacecraft,* 2017, PRRB Publishing, ISBN-1549651205.

Stakem, Patrick H. *Microprocessors in Space*, 2011, PRRB Publishing, ISBN-1520216343.

Stakem, Patrick H. Computer *Virtualization and the*

Cloud, 2013, PRRB Publishing, ISBN-152021636X.

Stakem, Patrick H. *What's the Worst That Could Happen? Bad Assumptions, Ignorance, Failures and Screw-ups in Engineering Projects, 2014,* PRRB Publishing, ISBN-1520207166.

Stakem, Patrick H. *Computer Architecture & Programming of the Intel x86 Family, 2013,* PRRB Publishing, ISBN-1520263724.

Stakem, Patrick H. *The Hardware and Software Architecture of the Transputer,* 2011,PRRB Publishing, ISBN-152020681X.

Stakem, Patrick H. *Mainframes, Computing on Big Iron,* 2015, PRRB Publishing, ISBN- 1520216459.

Stakem, Patrick H. *Spacecraft Control Centers,* 2015, PRRB Publishing, ISBN-1520200617.

Stakem, Patrick H. *Embedded in Space,* 2015, PRRB Publishing, ISBN-1520215916.

Stakem, Patrick H. *A Practitioner's Guide to RISC Microprocessor Architecture,* Wiley-Interscience, 1996, ISBN-0471130184.

Stakem, Patrick H. *Cubesat Engineering,* PRRB Publishing, 2017, ISBN-1520754019.

Stakem, Patrick H. *Cubesat Operations*, PRRB Publishing, 2017, ISBN-152076717X.

Stakem, Patrick H. *Interplanetary Cubesats*, PRRB Publishing, 2017, ISBN-1520766173 .

Stakem, Patrick H. Cubesat Constellations, Clusters, and Swarms, Stakem, PRRB Publishing, 2017, ISBN-1520767544.

Stakem, Patrick H. *Graphics Processing Units, an overview*, 2017, PRRB Publishing, ISBN-1520879695.

Stakem, Patrick H. *Intel Embedded and the Arduino-101, 2017,* PRRB Publishing, ISBN-1520879296.

Stakem, Patrick H. *Orbital Debris, the problem and the mitigation*, 2018, PRRB Publishing, ISBN-*1980466483*.

Stakem, Patrick H. *Manufacturing in Space*, 2018, PRRB Publishing, ISBN-1977076041.

Stakem, Patrick H. *NASA's Ships and Planes*, 2018, PRRB Publishing, ISBN-1977076823.

Stakem, Patrick H. *Space Tourism*, 2018, PRRB Publishing, ISBN-1977073506.

Stakem, Patrick H. *STEM – Data Storage and Communications*, 2018, PRRB Publishing, ISBN-1977073115.

Stakem, Patrick H. *In-Space Robotic Repair and Servicing*, 2018, PRRB Publishing, ISBN-1980478236.

Stakem, Patrick H. *Introducing Weather in the pre-K to 12 Curricula, A Resource Guide for Educators*, 2017, PRRB Publishing, ISBN-1980638241.

Stakem, Patrick H. *Introducing Astronomy in the pre-K to 12 Curricula, A Resource Guide for Educators*, 2017, PRRB Publishing, ISBN-198104065X.

Also available in a Brazilian Portuguese edition, ISBN-1983106127.

Stakem, Patrick H. *Deep Space Gateways, the Moon and Beyond*, 2017, PRRB Publishing, ISBN-1973465701.

Stakem, Patrick H. *Exploration of the Gas Giants, Space Missions to Jupiter, Saturn, Uranus, and Neptune*, PRRB Publishing, 2018, ISBN-9781717814500.

Stakem, Patrick H. *Crewed Spacecraft*, 2017, PRRB Publishing, ISBN-1549992406.

Stakem, Patrick H. *Rocketplanes to Space*, 2017, PRRB Publishing, ISBN-1549992589.

Stakem, Patrick H. *Crewed Space Stations,* 2017, PRRB Publishing, ISBN-1549992228.

Stakem, Patrick H. *Enviro-bots for STEM: Using Robotics in the pre-K to 12 Curricula, A Resource Guide for Educators,* 2017, PRRB Publishing, ISBN-1549656619.

Stakem, Patrick H. *STEM-Sat, Using Cubesats in the pre-K to 12 Curricula, A Resource Guide for Educators*, 2017, ISBN-1549656376.

Stakem, Patrick H. *Lunar Orbital Platform-Gateway*, 2018, PRRB Publishing, ISBN-1980498628.

Stakem, Patrick H. *Embedded GPU's*, 2018, PRRB Publishing, ISBN- 1980476497.

Stakem, Patrick H. *Mobile Cloud Robotics*, 2018, PRRB Publishing, ISBN- 1980488088.

Stakem, Patrick H. *Extreme Environment Embedded Systems,* 2017, PRRB Publishing, ISBN-1520215967.

Stakem, Patrick H. *What's the Worst, Volume-2*, 2018, ISBN-1981005579.

Stakem, Patrick H., *Spaceports*, 2018, ISBN-1981022287.

Stakem, Patrick H., *Space Launch Vehicles*, 2018, ISBN-1983071773.

Stakem, Patrick H. *Mars*, 2018, ISBN-1983116902.

Stakem, Patrick H. *X-86, 40th Anniversary ed*, 2018, ISBN-1983189405.

Stakem, Patrick H. *Lunar Orbital Platform-Gateway*, 2018, PRRB Publishing, ISBN-1980498628.

Stakem, Patrick H. *Space Weather*, 2018, ISBN-1723904023.

Stakem, Patrick H. *STEM-Engineering Process*, 2017, ISBN-1983196517.

Stakem, Patrick H. *Space Telescopes,* 2018, PRRB Publishing, ISBN-1728728568.

Stakem, Patrick H. *Exoplanets*, 2018, PRRB Publishing, ISBN-9781731385055.

Stakem, Patrick H. *Planetary Defense*, 2018, PRRB Publishing, ISBN-9781731001207.

Patrick H. Stakem *Exploration of the Asteroid Belt*, 2018, PRRB Publishing, ISBN-1731049846.

Patrick H. Stakem *Terraforming*, 2018, PRRB Publishing, ISBN-1790308100.

Patrick H. Stakem, *Martian Railroad,* 2019, PRRB Publishing, ISBN-1794488243.

Patrick H. Stakem, *Exoplanets,* 2019, PRRB Publishing, ISBN-1731385056.

Patrick H. Stakem, *Exploiting the Moon,* 2019, PRRB Publishing, ISBN-1091057850.

Patrick H. Stakem, *RISC-V, an Open Source Solution for Space Flight Computers,* 2019, PRRB Publishing, ISBN-1796434388.

Patrick H. Stakem, *Arm in Space*, 2019, PRRB Publishing, ISBN-9781099789137.

Patrick H. Stakem, *Extraterrestrial Life*, 2019, PRRB Publishing, ISBN-978-1072072188.

Patrick H. Stakem, *Space Command*, 2019, PRRB Publishing, ISBN-978-1693005398.

CubeRovers, A Synergy of Technologys, 2020, PRRB Publishing, ISBN-979-8651773138.

Robotic Exploration of the Icy moons of the Gas Giants. 2020, PRRB Publishing, ISBN- 979-8621431006

Hacking Cubesats, 2020, PRRB Publishing, ISBN-979-8623458964.

History & Future of Cubesats, PRRB Publishing, ISBN-979-8649179386.

Hacking Cubesats, Cybersecurity in Space, 2020, PRRB Publishing, ISBN-979-8623458964.

Powerships, Powerbarges, Floating Wind Farms: electricity when and where you need it, 2021, PRRB Publishing, ISBN-979-8716199477.

Hospital Ships, Trains, and Aircraft, 2020, PRRB Publishing, ISBN-979-8642944349.

2020/2021 Releases

CubeRovers, a Synergy of Technologys, 2020, ISBN-979-8651773138

Exploration of Lunar & Martian Lava Tubes by Cube-X, ISBN-979-8621435325.

Robotic Exploration of the Icy moons of the Gas Giants, ISBN- 979-8621431006.

History & Future of Cubesats, ISBN-978-1986536356.

Robotic Exploration of the Icy Moons of the Ice Giants, by Swarms of Cubesats, ISBN-979-8621431006.

Swarm Robotics, ISBN-979-8534505948.

Introduction to Electric Power Systems, ISBN-979-8519208727.

Centros de Control: Operaciones en Satélites del Estándar CubeSat (Spanish Edition), 2021, ISBN-979-8510113068.

Exploration of Venus, 2022, ISBN-979-8484416110.

Patrick H. Stakem, *The Search for Extraterrestial Life,* 2019, PRRB Publishing, ISBN-1072072181.

The Artemis Missions, Return to the Moon, and on to Mars, 2021, ISBN-979-8490532361.

James Webb Space Telescope. A New Era in Astronomy, 2021, ISBN-979-8773857969.

Riverine Ironclads, Submarines, and Aircraft Carriers of the American Civil War, 2019, ISBN- 978-1089379287.